Learn C# in 21 Days:

A Complete Schedule to Teach Yourself C# in 21 Days

Griffin Gallant

Table Of Contents

Introduction

Although there is nothing easy about learning a new language of any sort, the aim of this book is to allow you to pick up and understand as quickly and easily as possible. While doing so, my goal is to provide you with complete and comprehensive skills that are able to help you retain important knowledge and guide your programming and coding experience in C#. In order to allow you to learn in the most efficient manner, I urge you to do the practice following each lesson that will reinforce the lessons we explore in this book.

As the title of this book suggests, within 21 days I want you to have a firm grasp on C# as a language. For this purpose, I have broken down the process of learning into 5 chapters that first cover a basic description of what C# is, then go through each week of learning and finally give you some helpful hints and resources to reinforce any subjects or skills that you may need a little more assistance with. Within each weekly chapter, I have broken down your work into an explanation and description of one portion of the C# programming and coding language structure or component.

This language is one that will mirror some components of other computer programming languages—it is almost invaluable experience if you are familiar with other languages themselves. Although this book is not quite just a beginners manual that introduces you to the art of programming, it will reiterate and acknowledge a lot of these skills in an effort to help you achieve your maximum programming results in the quickest amount of time you can. Keep in mind that this language is one that provides the programmer with many abilities, this book may refer to things in terms of gameplay programming I hope you are also familiar with.

Although I provide you with the information and some basic skills to practice, it is up to you to put these skills to use. I highly recommend you find a strict routine to abide by in the 21 days that involves research, study of the material and practice in your own virtual environment that familiarizes you with these skills. C# can seem like an overwhelming task to teach yourself an entirely new skill. I can give you the knowledge and resources but it will be up to you to take on the challenge of synthesizing and using them to create your own programming and codes.

With an object oriented and structured language such as C#, I think you will find it easy to learn and user friendly. This high level language builds on and, in some ways, complements the languages of C, C++ and even Java. However, even if you are a master of these languages, this new task will have just a steep of a learning curve as any of the other programming languages do! Throughout this process, I encourage you to use your own knowledge and instincts to apply the concepts and strategies in this book. Remember, everyone is a beginner before they are a master!

Chapter 1- What is C#?

You have come across and purchased this book in order to learn C# programming. Throughout this book, you will learn basic rules of how to apply C# within your own programming projects. C# is a computer programming language that was both designed and developed by Microsoft. This is a fairly new language and is considered to be a language that is very functional and object oriented in nature. Although C# is most certainly not the same language as C or C++, there are definitely some similarities between the languages and it could even be considered helpful to have an insight or knowledge and experience in either or both of those languages. Just as with all programming languages, having knowledge of how they work on a basic and functional level is helpful. All languages have one thing in common, they are working towards telling the computer what and how to set up systems, databases or how and when to run an execute programs.

This particular language has roots stemming from a programmer, Anders Hejlsberg, who was a member of the development team of .Net Framework for Microsoft. Although some people tend to criticize C# as being somewhat of a "rip off" of Java, you will see throughout the course of this learning manual that (if you have a knowledge of Java that is) there are distinct differences in the content and structure of the language that make it different. With a design that was formulated for an environment that has the ability to use high level languages and consists of executable code (also referred to as Common Language Infrastructure or CLI) it appeals to many who are interested in this easy to learn language.

With a language that boasts efficiency as one of its objects, there are quite a few advantages to using it. Throughout this book, I will introduce you to how this object and component oriented language can be easily acquired as a new language for you to use. Along with being a structured language, you can count on efficient results from your programing and coding within the C# language. Throughout the chapters of this book, I will assist you with optimizing your performance, in order to obtain your desired output from the work you put into programing and coding in this language.

Although C# was developed by using the Microsoft platform, the language itself is not just limited to it. Another great concept of C# is that this versatility allows you to take the object oriented language to different platforms to create multi-platform apps, programs, and software. As I am sure you are familiar with other languages, certain aspects of languages benefit certain desired results. There are many benefits, that will be covered in this 21 day guide to learning C#, that assist a user with portability, typing, metaprogramming, property, memory access and methods and functions—among an array of other positive attributes we will discuss and show examples of in the coming chapters.

This language has only been around, in official versions, since 2002 and therefore could be considered a newer programming language. However, do not let that lead to you believe it cannot and does not compete with other languages, such as Java, that have been around a little longer. Developers of this language really aimed to improve upon and add different features, initially designed specifically for the .Net framework but now have been expanded, and were successful in this with many features—one many tend to like is the Syntax used.

One thing I would like to bring up is the possibility for this language to be used in order to focus on gameplay. It is important to gamers that, not only is the game interesting and holds one's attention, they also want the experience itself to be enjoyable. As you are probably somewhat of a programmer yourself (or at least appreciate the capabilities of computers) after reading this book I expect you will have a great understanding of just how hard gaming programmers work in order to create the experience you have as an end result.

Although I do not think reading this book alone will allow you to create a masterful game (you need a lot of practice as well!)—I do believe that you will understand concepts to a greater degree that are essential when you are dealing with the many components of graphics programming and design. Keep in mind, throughout the book, that you are creating and fine tuning a potentially endless world through your coding and programming skills in this language. The object oriented component really helps to reinforce this idea.

In the coming chapters you will learn both how and why C# will be very useful to you and how you can easily pick up and become fluent in all the basics within less than a month. With a standard library that rivals any of the other languages, by the end of this book you will know how to find the source code you may need and execute it to produce fairly consistent and exquisite results. Another positive feature of this language is the Boolean conditions, which are favored by many, along with the Windows integration that allows you to access one of the most popular operating systems with ease.

As a final note, I just want to point out a few general concepts of programming and coding that are good to keep in mind throughout this process. First of all, remember that computers do not think. This means that they will execute tasks exactly as you tell them. A computer will not infer a message from your code, as a human might, always keep this in mind when you are having any struggle. Also remember that a simple mistake in line endings or a wrong letter can cause a whole lot of headaches. If you come to, what may seem like, an impossible impasse—perhaps you need to take a quick break and come back to look at your coding again. Sometimes we cannot see our own mistakes when we have been staring at a screen for a long period of time. Finally, remember that you are not the first person to have a struggle with this language and will certainly not be the last. The last chapter of this book will be a guide to further helpful resources and hints as to address any struggles I foresee you having in the future as you begin to learn this language.

This book sets up reasonable expectations for you to learn C# within a three week, or 21 day, period. I encourage you to try to do learn in that exact manner. It is important for you to not overwhelm yourself with too much content in a short period of time. Conversely, don't allow too much time to lapse before coming back to your work. You may forget subtle nuances that may drive you crazy. Allow yourself to have reasonable expectations, which this book is designed to do. Lastly, enjoy yourself! Learning a new language should be an enjoyable and fun experience for you. Allow yourself to take pride in the new skill you are learning and get ready for a crash course in basic C# mastery!

Chapter 2- Week 1

This first week of learning we want to focus on building a strong foundation that gives you a full picture of the basics of C#. Essentially, by the end of week one you will have a firm grasp on the environment in C#, the program structure, basic syntax and data types, type conversion, variables, constants and operators. Within this section, there will not be as many exercises as there may be throughout other sections; however, I will provide you with a firm explanation and example or terminology that will be useful for you throughout your first week. As I stated in the last chapter, if you are unfamiliar with the terminology or methodology used, I will try to give a quick, basic review of the terms. If this does not satisfy jogging your memory, you may need to find a quick refresher from Google or another resource.

I would also like to recommend a free interactive shell program you can try out code on—csharppad.com, this should help you complete any training exercises included in the book.

Day 1- Environment

In the first chapter we discussed a bit about how C# was designed to be integrated with Microsoft and .Net Framework by the developers. As one would guess, this is a large influence on the environment you find within C#. An important part of understanding C# comes from the understanding you have of how the two interact with each other. First, let me give you a very brief synopsis of .Net framework to help you better understand how the two interact.

As the latter part of the name framework suggests, the .Net framework is a platform in which one can run applications. There are three different types of applications that the framework can assist you with writing—web services, web applications and windows applications. When one is using the framework applications, it can actually be used by a variety of programming languages, as it is a multi-platform application. The multitude of languages, from Visual Basic, Javascript, C++ and COBOL, also have the ability to communicate with each other within the framework. Another great attribute of this framework is the extensive code library that can be used by a language, for our purposes we will be talking about C#.

One of the great things about any language or framework is the potential for a library of code you can utilize. For the .Net framework and C#, there is a huge library that you can access and utilize for your coding projects. One advisory statement I always give people is that, although these libraries are excellent resources, be sure you understand the how's and why's of the coding itself to insure that you are able to fix an issue, should it arise.

Microsoft itself also offers a great tool for its users in order to assist their development in their C# programming projects. There are two excellent ones that the company offers for free, great resources for writing any kind of application, from simple to complex. The name of the two I would recommend are Visual C# 2010 Express and Visual

Web Developer. If these options are not something you feel comfortable with or enjoy using, you can always use a text editor in order to write the source code you may need. I recommend the other options because I find them to be a step up in usefulness from a simple text editor. However, lots of people prefer that method—it's really up to you and the comfort level you feel with a program. Test them out and see which methodology works better for you. However keep in mind you will be missing a huge asset in the intellisense if you use a plain text editor. You can download the two Microsoft programs I spoke up from a website called Microsoft Visual Studio.

Finally, you can also run C# on Linux or Mac OS, which is a great ability to code across different platforms. However, in order to so do, you obviously need a non Microsoft framework to do so on your different operating system. I recommend "mono"—which is an open source option for use on both Linux and Mac.

Now that you have a basic idea of the environment, go download or open the programs you would like to get familiar with to start your coding process. Play around with the software—there are tons of tutorials out there about how to set these environments up out there if you need help, once you're done familiarizing yourself with the software you are ready for day 2!

Day 2- Program Structure and Basic Syntax

Now that we have established a little about how the environment was created for C#, let's look at the actual structure and how you will interact with programs through the syntax of C#. When we are looking at how C# is structured, we are looking at 7 basic components that make up the program. These are that there are class methods and attributes, namespace declaration, a class, a main method, comments and statements and expressions. Now, what do these look like for you in terms of your programming and coding? Why don't we take this "Hello World" example and break it down into parts to see this in action.

```
Using System:

Namespace HelloWorldApplication

{

   class HelloWorld
```

```
{
    static void Main(string[ ] args )
    {
        /* my first program in C# */
        Console.WriteLine("Hello World");
        Console.ReadKey( ) ;
    }
}
```

So hopefully this looks somewhat familiar to you and, although you may not quite understand the meaning, you know all the symbols and spaces are important. When we are using C#, the first line is always going to consist of using System, although most of the time you will see more than one statement with the first key word being using. The next line in C# will be used as a declaration of the namespace. In case you are unfamiliar, we should think of the terminology of namespace as a descriptor of the specific group or collection of classes. In our Hello World example, we are looking as that phrase as the class (HelloWorld). Following the namespace line, you will come to the class declaration. The class is going to inform you of the method of the class. It is a typical characteristic of methods that there are multiple within one singular class. It is important to be familiar with the methods of a given class because you can then be more aware of the behavior of the class you are dealing with. In the Hello World example, we are looking only at one single term, or main method.

Each class has a main method that you can find in the line following the class declaration. Essentially, when you are looking at this line you can expect the main method to tell you what is going to happen when the code is executed. This point is often also referred to as the entry point in C# programs. When you see the /*...*/ in the following line, you are looking at code that will be ignored by the actual compiler and instead is added in the program under the comments.

The last two phrases we are looking at in this example are Console.WriteLine("Hello World") and Console.ReadKey(). These lines are telling the system to both display your message and controls how and when the program runs and closes. These are obviously both important parts of how and why the program would work and function and you want to make sure of a couple of key components that could affect your coding, just due to similar grammatical type errors you may find in the English language.

You want to make sure that you note a subtle difference (if you are familiar with Java) that the class name and program file name can be different. Also you want to make a note that C# is both case sensitive and, rather than ending sentences with a period or other notation, they are ended using a semicolon. Also always make sure to note that the main method is the point at which the program execution starts. These four components are good to note in order to avoid annoying and tedious problem solving when mistakes occur with your coding.

Play around with the above example within the virtual environment I suggested on the first day. Now we will discuss a little about the syntax of C# and how some of the concepts we have talked about with the structure really develop and exemplify the syntax. For starters, the /*...*/ that allows you to comment on a program is an important feature to remember. In addition to this, it is good to remember that when naming a class, the first character in the name can only be a letter, not a numerical digit. Although you can make up the class itself following the initial digit with letters and numbers, always remember this and that symbols are always not allowable in the naming of a class. Finally, the name cannot contain a C# keyword. There are two different categories of keywords, contextual and reserved, below I will list them all in their respective categories.

Contextual Keywords—get, partial, from orderby, let, dynamic, set, join, descending, ascending, into select, alias, group, remove, global and add.

Reserved keywords—case, decimal, event, for, int, new, continue, enum, in, float, byte, namespace, in, long, else, fixed, const, break, finally, lock, implicit, double, class, is, if, bool, do, false, checked, base, goto, internal, extern, delegate, as, char, foreach, explicit, interface, default, abstract, catch, try, internal, is, interface, sealed, null, switch, object, short, protected, this, operator, sizeof, is, public, throw, out, readonly, lock, true, long, red, out, static, override, namespace, string, return, typeof, params, new, struct, uint, sbyte, volatile, unchecked, throw, stackalloc, ref, private, while, void, virtual and sealed.

Day 3- Data Types

Now let's talk about data and the different types you will see in C#, which we have three basic categories of. These categories are value, pointer and reference types. If you are familiar with programming, you are definitely familiar with the different types of values that can be associated with data types.

Values types of data are, as the name would suggest, data types that consist of the value themselves. These are terms like Boolean data, byte or int. A good test to run within your virtual environment is to get the size of a given data type using the term sizeof(type), only replace type with the data type you wish to know the size of. A complete list of the value data types are decimal, double, bool, float, char, byte, long, sbyte, short, uint, int, ushort and ulong.

Within the reference type of data, we are looking at (you probably guessed it!) a reference to a type of data. These types of data are able to tell you where memory is located. There can even be multiple references to multiple variables. His type of data, we are looking at string, dynamic and object as our key words to describe the reference data types.

The last type of data, pointer, points to a different variable in order to direct you to the appropriate location. In order to reinforce that you do indeed know what a pointer is, here is a generic sample of what a declaration of a pointer variable should look like–

type *var-name;

Go ahead and play around with the variables in your own dictionary and see what kind of output you can get for these. There is one other important phrase I think you should be aware of in your programming and coding. This is when you use the /unsafe command–

csc /unsafe darkprog.cs

In the above example, the program we would be specifying as unsafe is “darkprog.cs.

Day 4- Type Conversion

As the word conversion would suggest, a type conversion is changing or converting a given type of data into a different one. Sometimes you will also see this referred to as a type conversion. Within C#, there are two distinct types of type conversions. There two types are known as the implicit type conversion and the explicit type conversion.

The implicit type conversion are performed to do actions like conversions that are from a base class to a derived class. These are performed in what is referred to as a “type safe” manner. The other type of conversion, explicit, require a cast operator in order to complete their function. There are 16 different functions within C# that are able to be used as a built in type of the conversion methods. I am going to give you a list of examples that you should practice becoming familiar with within the C# environment.

- ToChar- allows you to take a type and convert it to a single Unicode character
- ToByte- allows you to get a byte by converting it from a type

- ToDateTime- allows you to get a date time structure by converting a type (either string or integer)
- ToBoolean- allows you to get a Boolean value from converting a type
- ToDouble- allows you to get a double type from converting a type
- ToDecimal- allows you to get either an integer or decimal type converted from a floating point
- ToInt64- allows a 64 bit integer to be converted from a type
- ToInt32- allows a 32 bit integer to be converted from a type
- ToInt16- allows a 16 bit integer to be converted from a type
- ToSbyte- allows a signed byte type to be converted from a type
- ToString- allows a string to be converted from a type
- ToSingle- allows a small, floating point number to be converted from a type
- ToUInt64- allows an unsigned bit integer to be converted from a type
- ToUInt32- allows an unsigned long type to be converted from a type
- ToUInt16- allows an unsigned int type to be converted from a type

- ToType- allows a specified type to be converted from a type

For your exercise today, I encourage you to go find programming that does each and every one of these functions/conversions in order to get more familiar with what they involve and how they can assist you in your own programming. Remember that although your coding may be long, your output will be short—but should always produce the answer you were trying to achieve from the coding in question that you have done.

Day 5- Variables

Now that we have discussed how to classify coding and data into classes, how to use syntax to communicate with the program and different types of conversions you can complete with coding—let's talk about the variables that allow you to store these in. That is, in essence, what a variable is—something that allows you to store values within memory and how you would like them to be executed (or the set of operations). There are five different types of variables that you will be utilizing within your programming and coding in C#, these are—boolean types, nullable types, Floating point types, decimal types and integral types.

Within each of these types, certain data types that we discussed previously are the examples of what is stored within a given variable.

- Boolean types— I am sure you are familiar with what a Boolean variable is. However, easily put, this is where your true and false values are stored.
- Nullable types— This is a fairly easy one to remember as well. Within nullable variables, only nullable data types are stored!
- Floating point types— within this variable you will see both double and floating data types being stored
- Decimal Types—simply put, this variable only stored the decimal type of data. This one is also pretty easy to remember!
- Integral types— within the integral variable, you will see many types of data stored. These are long, ulong, char, sbyte, byte, short, ushort, uint and int (by far this variable has the largest amount of differing data types that can be stored within it)

Now let's go over a little of the syntax for different ways of summoning variable characteristics.

Practice for defining variables by using this general variable definition—

<data_type> <variable_list>;

Try to always remember the key to initialization when you are working with variables in this context. Which is that you should always, always practice doing them in the correct manner. If not, your program may produce results that you are not wanting or expecting. How about for the initialization of variables equation that can be used for general purposes? This looks like–

variable_name = value;

or, for declarative purposes–

<data_type> <variable_name> = value;

How about when you want to have the ability to accept certain input and be able to store it in a given variable. If we look at the System namespace and the Console class, which provides a function ReadLine(). Here is an example–

int num

num = Convert.ToInt16(Console.ReadLine());

The last thing we will discuss for today are the two different types of expressions used in C#. These are known as lvalue and rvalue. These expressions tell us where each may appear within an assignment. The lvalue can be seen on either the left or right handed side. However, rvalue can only appear on the right handed side of the given assignment. It will never be seen in the left handed side of the assignment. This should be food for thought and something that you allow yourself to play with before moving on to the next days' work.

DAY 6- CONSTANTS

When you think of constant, you think of something that remains the same no matter what. This is very much what the constant, fixed values within a program represent as they are unalterable when executing. Another commonly used name for these constants is literals and they can be made up of any of the data types. For example we can see string literal or a floating constant. As a constant is a variable, it is treated in the exact same way as any other one—except for the fact that it cannot be modified after they are defined. Let us look at the types of constants and the examples for using them within your own programming and coding.

With integer literals, the constant we see is either hexadecimal or a decimal constant. There is no prefix id for decimal; however there is one for hexadecimal—0x or 0X are the two prefixes you may see. As for a suffix, this can be either L or U for long or unsigned and can be upper or lowercase and in any given order.

Hexadecimal—0x5a /*hexadecimal*/

Decimal—65 /*decimal*/

Int—20 /*int*/

Unsigned int—20u /* unsigned int*/

Long—20l /*long*/

Unsigned long—20ul /*unsigned long*/

Floating point literals are those that have a point, that represents a decimal point or a part of the given integer and can be represented in either exponential form or decimal form. Here are some examples—

4.5216 is /* Legal */

45216E-5L is also /* Legal */

However, .e22 or 810E or 640f are all illegal and incomplete forms. Make sure to play around with these if you aren't sure which format is correct within the program.

Within C#, there are character literals or constants that are referred to as escape sequence codes. I am going to give you some examples and I would like you to go and figure out what their meaning is within C#. These will be surrounded by quotations but only use the characters within them, remember they all begin with a /. Here they are "/f", "/n",

"/b", "/a", "/""", "/?", "/'", "/t", "/r", "/v" and finally, "/xhh". Go ahead and try these out and see what their features are within the program.

As above I was using " " in order to show you examples of each of the different constants, string literals will actually use and be enclosed by these double quotes. Let me show you a few different examples, remember that the quotes around the characters are what define the string constants. All of the following will actually represent the same things, just using a different method.

"welcome, son"

"welcome, /

son:"

"welcome, " "s" "on"

@"welcome son"

The last thing I want to cover is a simple way in order to define a constant. The keyword you want to remember with this is very intuitive "const". Let's look at the general syntax you will use when you want to use this function—

const <data_type> <constant_name> = value

Day 7- Operators

What does the operator on the telephone allow you to do when you call them? They connect you to another given series of numbers (we call a phone number) in order to get in touch with another person. Within C#, operators are the symbol that triggers the program to perform a given manipulation. The type of operators we see in C# are arithmetic, relational, logical, bitwise, assignment and misc.

As the name arithmetic would suggest, these operators are those that carry out functions of basic arithmetic. These are going to utilize symbols like ++, --, %, +, *, - and /. All of these carry out the basic functions you use on your computer's calculator like adding, subtracting, multiplying and dividing.

These next few symbols are those that tell of a relational function being carried out by the operator. These symbols are ==, !=, >, <, >= and <=. Play around with these as well by, once again, putting two different integers both in front of and following the given symbol.

These next operators fall into the category of logical operators. These consist of &&, ||, !. The last one in this list, !, is something that is not intuitive for you to figure out. You would actually place this in front of one of the other two in order to define it as a logical NOT operator. Try these out as well and see what they output.

As the name bitwise operator would suggest, these types of operators are made up of bits. They speak to the binary within the system and tell it what to do using similar signs, but long series of 0s and 1s, as is the language of binary. You can play with this method a bit with different signs and mathematical symbols we have looked at previously, just using an equal sign to equal a binary number. This would be something like A>> 2= 15 (this would right shift the operator). Or even simply (A=B)= 12—translated would be 0000 1100.

Assignment operators consist of the mathematical symbols we notated along with a = in order to make an and/or statement. Something like X= Y+Z would assign a value of Y+Z to X. You can use any of the mathematical symbols in order to make statements with such an operator.

Finally, misc or miscellaneous operators are those that carry out other functions that do not fall into the categories above. These are sizeof(), typeof(), & and words like is or as.

Now you have a great deal of different exercises to try out and master—best of all, you have completed your first week of learning! By this point, especially if you have other programming and coding language, you should begin to feel confident in the basics of the language that allow you to perform basic tasks (later building on these to create more complex coding and programming). Great work for your first week—now on to the second!

Chapter 3- Week 2

Welcome to your second week of C# learning and training! This week we are going to take some concepts and look at them with a little more of a critical eye. In order to take this next step, please make sure you feel confident that you understand how to utilize the skills we talked about in week one. You are most likely not feeling as though you have mastered the language, I do want you to feel like you are an experienced user of the previous week's tools.

Day 1- Decision Making and Loops

Within the world of C#, and programming and coding in general, decision making is essentially a mirror image of the basic format of real life. Within C#, and programming in general, you are asking the program to test or evaluate one or more conditions. Along with these conditions, there is a statement(s) that are completed or executed if the condition is met.

Within C#, there are 5 different statements that are used for decision making purposes. These are if, if/else, nested if, switch and nested switch statements. These statements are probably pretty familiar to you if you have ever used another programming and coding language. If statements are Boolean in nature, with only a true or false answer. If/else statements have an optional statement that may run when the Boolean expression is proven to be false. Nested if statements are if statements that are within another if statement. Switch statements are those that test a list of values against a given variable in order to determine equality. Finally, nested switch statements are (you probably know or guessed!) one switch statement that is nested within another switch statement.

Here is a quick example of what an if/else statement, also could be called a conditional operator (from the previous chapter) would look like in a generic manner.

Try1 ? Try2 : Try3;

Play around with different terms and variables from the previous week within the context of these statements. See what you can put inside of other statements in order to get your desired results.

Loops take the idea of statements that need to be executed and add the additional element of sequential execution of code. The loop is where you can start to really make a very simple mistake and, if you are unaware of what proper code should look like, this mistake could cause you a Mount Everest size headache. Just as with the decision making we just discussed, loops allow for a few options of how they can be executed and carried out.

These four options are while, for, do/while and nested loops. While loops are those statement or statements that test a given condition, determine it to be true and then execute the body loop. For loops are those that continue to execute statements in a given order and then abbreviate code to manage the loop more efficiently. The do/while loop statement is similar to the while loop, however instead of testing the condition at the start to allow the execution, it is tested at the end. Last the nested loop if one that can be looped within the confines of another loop. If you wanted to create something called an infinite loop, defined just as the name would suggest, something that continues forever! You would want to use the for loop that leaves conditional expression empty in order for it to continue to continuously run.

We can also place two types of conditions on loop statements, break and continue statements. As they names suggests, break statement transfers execution that comes directly after the switch; whereas continue immediately and automatically allows for a retest before it continues to loop.

Test out loops within your code and build on the skills you have covered this far for today. Hopefully looping is a fairly familiar concept that you will easily pick up on, if not, do some more studying of the terminology and then try again!

Day 2- Encapsulation

As the intuitive nature within yourself has lead you to believe, or know, that encapsulation is literally enclosing things inside of each other, in this case preventing access to details of how things are implemented. Within this day of learning, we are talking about how to use an access specifier in order to apply a certain access to a class member in either private, public, protected, internal or protected internal context.

As public suggests, a public access specifier is one that is something that can be accessed or exposed to other objects and functions within your programing. Private functions can only be accessed by other functions or objects that are also private. Protected access will actually be discussed in next week's lesson on inheritance, we just mention it here to bring attention to it so you are aware it exists and will be addressed later. With internal access, we are looking at the exposure of both member functions and variables to objects and functions that lie within the current class or method of the given application. The protected internal access will also be discussed further in the inheritance chapter for you in next week's material.

Within larger sections of coding of various different methods, statements and variables, you are going to use encapsulation in order to define terms for access. For example, you would use "class rectangle" followed by specificities on the code you would want executed, and end with //end class Rectangle. Within that code you would also define member access starting in the second line of your code.

DAY 3- METHODS

Methods are the ways in which you tell or declare the structure elements you are going to use. This will look like the following—

```
<Access Specifier> <Return Type> <Method Name>(Parameter List)

{

    Method Body

}
```

Now let me define each of these methods for you so you can use them in order to try out different types of methods within the generalized code. The five different types are access specifier, return type, method name, parameter list and method body. When you want to see the visibility or variable of a method from a different class, you will use an access specifier. A return type, as the name suggests, will return a value. If there is no method to be returned, it is known as the method type “void”. Something that cannot be used by any other identifier, and is unique to a method, is called the method name. The parameter list will be an enclosed list of parameters that are used to receive and pass data from things like type, order or number that define a method. Essentially, laying out the how and whys (parameters) or how the method operates. Finally the method body is simply put the complete instructions that are used to execute a given activity. Try these out with the generalized format above and see how you can create firm definitions within yourself of each of these.

To note, parameters can be passed through each other in three different ways–value, reference or output. These names are named after the manner in which the pass through and should make sense on the basis of their names alone!

Day 4–Nullables and Arrays

Nullables are those things that you assign values too both within normal and null ranges. Let me give you the general example of how you are going to assign these null values—

< data_type> ? <variable_name> =null;

In tandem with nullables, you are going to use what is called a null coalescing operator the reference type and nullable value. You want to use this when you have a very exact and implicit conversion that you want to carry out. You are going to create code that addresses the multiple null variables as well as define them. Remember that you are going to want to place all brackets and spacers properly, along with the proper capitalization!

Where nullables are things that are being tested upon in a given circumstance or order, arrays store same elements within fixed and sequential collections. Basically, instead of declaring things individually, you are going to declare them as whole with contiguous, boxed type memory locations. The smallest or lowest address will be associated with the first element; whereas the highest will correspond with the last. When declaring arrays, syntax will look like this –

datatype[] arrayName;

Within this example, in order to specifically define the elements within the array, you use the term datatype. The [] then are used for you to tell the rank of the given data type within the array. Finally, as you might surmise, you will place the arrayName where the term is in the equation.

The types of arrays are multidimensional, jagged, passing arrays to functions, param arrays and the array class. Multi-dimensional arrays are those that are the simplest type and can consist of simple two dimensional arrays (or more). Jagged are arrays that are multidimensional and also are arrays of other arrays. Passing an array to a function is where you are specifying an array but do not have an index. Param arrays are those that are used when there are many parameters passing through a given function, it is also unknown how many will be passing. Finally, array class is really the basic and basis for all arrays—allowing for different formulations and manners in which the different arrays work together.

Day 5- Strings

We briefly mentioned strings earlier and, if you are familiar with programming, I am sure you have heard the terminology before. When you are stringing in the world of programming, you are putting together characters (or stringing them) in order to possibly create and declare a string variable you have put together using a string keyword. When you say string keyword, you are actually referring to the System.String class, just using a different word in order to do so.

Why would you want to create something using Strings? Let us look at the different methods you can use in order to create a string object. You will want to do so by utilizing a string concatenation operator, assigning a string variable to a string literal, getting a calling or property that enables a method returning a string, utilizing a string class constructor or finally by creating a string representation by converting a value or object—or calling a formatting method—to do so. You will want to exercise this by creating code that has structure like the following—

```
namespace StringApplication
{
   class Program
   {
      static void Main(string[] args)
      {
         //from string literal and string concatenation
         String fname, lname;
        fname = "Jane";
        lname = "Smith";
```

```
string fullname = fname +lname;

Console.WriteLine("Full Name: {o}", fullname);


//by using string constructor

Char [] letters = { 'H', 'i'};

 string greeting = new string(letters)'

Console. WriteLine("Welcome; {o}", welcome);


//methods returning string

string[] sarray = {"Welcome", "To", "The",
"Example" ];

string message = String.Join(" ", sarray);

Console.WriteLine("Message: {o}", message);
```

```
        //formatting method to convert a value

        DateTime waiting = new DateTime(2016, 11, 11,
17, 58, 1);

        string chat = String.Format("Message sent at {0:t}
on {0:D}, waiting);

        Console.WriteLine("Message: {0}", chat);

      }

   }

}
```

Practice and play with the above example until you are familiar with what the components are and how you can change them to create different outputs from the code. This is an excellent way to practice a more complex example of bringing to live code.

DAY 6- STRUCTURE

When you want to create a structure in C#, you will want to always remember to use the keyword struct in order to do so. A structure is what allows you to have a single variable and then have it hold a multitude of different data types that are related within it. Within the structure itself, you will want to be able to define and have clear ideas of what it should look like. When you have an object oriented language such as C# that creates great graphical interfaces, you will want to have a clear ideas of the boundaries of the universe you wish to create within a given structure.

You want to be able to create a structure that enables you to define simple and complex structures within your programming. First I want to share with you features of the structures and about how the compare and match up to classes. One interesting feature of structures is that they always have defined constructors but are unable to have destructors. You also have constructors that are unchangeable and clearly defined from the get go. Different from the classes that can inherit both other classes as well as structures—a structure is unable to produce and have this feature. You can, however, use a structure in order to put into place either a single interface—or more. Structures also have clear boundaries that are not able to be placed in categories like abstract, protected or virtual.

As for a few key differences, unlike classes structures are unable to have default constructors or support inheritance. Classes are what is known as "reference" types and structs are differing as "value" types. Using these key terms in order to give you a better understanding of how and why structures and classes differ you should be able to structure your coding and programming according to the output you wish to make.

Day 7- Enums

As with the previous sections shortened keyword, enums comes from the source word of enumerated. When it comes to this type of value data type, we are looking at enumeration as somewhat of a package of integers that are constants. One factor of an enum is that is it unable to inherit and strictly must consist of its own values. Here is the syntax we will use to generalize how to declare enumeration—

```
enum <enum_name>
{
    Enumeration list
};
```

The terms in the above generalized syntax refer to the list of identifiers (enumeration list)—please note that these values or identifiers must be separated by commas. As you probably could surmise, the enum_name refers to the type of enumeration name. Symbols within the list are values that are integers and they also must grow greater with each passing symbol, it cannot be the opposite way. You can also have lists that you enumerate that include things like days of the week that could be broken down further into week days and weekends. Enumeration exercises are a great way to end this week of your learning as they tie together some loose ends on how to bring together projects will additional details.

As we wrap up the second week of your learning of C#, I want to remind you of a few quick things, along with some description of the term enums. As you have now reached the point where you are almost 2/3rds complete with your learning of the C# language, let us take a moment to discuss how you see it playing in with the larger picture of coding and programming as so far described. This language can be used for many purposes, however, as you can see from the intricate features that allow for distinct detail, you can see how this language does lend itself well to graphical interface based projects.

When you think about a game that you play, you have many different aspects interacting in order to create the images you see on the screen. You have things like background images, your character, sounds and movements that are all occurring at the same time. With enough practice of this language and building of your skills, this will be something you can achieve for yourself with enough hard work and effort.

Chapter 4 - Week 3

You are now officially two thirds complete with your guide to C# programming in 21 days! By this point, I want you to feel as though you have mastered the first weeks content to the point where you can understand the basic operators, variables, constants and other concepts that make up the environment and syntax of C#. Hopefully by now you have a fairly good idea and understanding of the importance your strict attention to detail has on the ability to properly execute your coding and programming. The general guidelines of how you will structure your code should be something that feels good to you, although by no means should you expect yourself to be a master! Keep in mind that you are almost to the end of the learning process to allow yourself a strong foundation you need to succeed as a C#

programmer and coder. In this last chapter I challenge you to try to go back to a lesson from each previous week in order to just refresh your memory due to the massive amount of information you have learned within such a short amount of time.

Last week we focused more on in depth looks at some of the terms we mentioned in the first week and expanded on them to reaffirm and build on the concepts. In this final week, you will be putting the pieces together and giving yourself a clear understanding of why and how you will coding and programming after you have completed this book!

Day 1- Classes

We talked about structure a bit last week and even addressed some of the differences between structure and class. We are starting this week out with a firm definition, description and some tools and examples for you to utilize as your understand class in terms of C#. Classes are essentially outlines that actually could be used to define the term object (in the object oriented language of C#) in terms of a class itself. Variables and methods actually are members of the class and you are using classes in order to have an idea and description of what you will find inside of a class.

When you start to define a class within your programming, you are actually going take into account member variables within the object you are working within. You want to always take into account that member variables are only able to be used and accessed by member functions. This is just one example of why a class may be an important factor in the decisions you make with different code. You always want to take into account you audience and the work and effort you have put in to different class members.

With constructors of class you always want to remind yourself that these are important, as they are always executed when new objects are created within a class. The constructor will not have a return type—however it will hold the same name as the class. This is important for you to both remember as you code and program in an effort to potentially track down a potential mistake if they do not line up properly. We also need to remember that some constructors are also parameterized and some are default. You need to be able to know that if you need to place parameters on your class you are dealing with, do not simply go for the default constructor! It is important for you to set up parameters of success for yourself.

Destructors are the last part of the class that you should never forget about—as they are executed at the end of an object. The only difference between the name of the class and the destructor will always be the tilde (~) and also remember that there will never be any parameters associated with it. They also are without the ability to be overloaded or inherited.

Day 2- Inheritance

We have not referred to inheritance within many different sections of this manual without fully explaining its importance or what exactly it does in relation to these vague terms. Just as you would associate an inheritance from a relative, there are many instances where the parameters or input/output instructions are inherited by different aspects of the code you create. The best way to look out and find these instances is too look for places where you get a result you didn't quite expect, due to it seemingly having similarities to previous code loops or variables that may have come up within your work.

Remember that you can inherit these from a base class or even a derived one. It can also happen from multiple interfaces or base classes, sometimes causing something to happen that you may not want to. Generally, there is an easy fix for something like this and has to do with a small detail you would need to fix to disallow the inheritance.

In order to play around and learn more about how this happens, I encourage you to take the lessons from the previous chapters and find places that you can manipulate and change them, look at the resulting output and see an example come to light right before your eyes. As with all the practices in this book, it is always better to practice them for yourself.

Day 3 – Polymorphism and Operator Overloading

As the root "poly" of polymorphism suggests, there are multiple forms that you can use within this language that allow it to have diverse and far reaching programming potential. A polymorphic interface is one that you can have multiple functions that come from a singular interface—broken up further into static or dynamic types. Within the static polymorphism, you are looking at a compile time of a function that occurs in a response to this occurrence. Dynamic polymorphism, on the other hand, does not have the response happening when something happens, it is predetermined from the start of run time.

Static or early binding that links compile time with a function in static polymorphism has two techniques that assist with this implementation. These are operator and function overloading. A good example of what occurs when a function is overloaded is when someone tries to print something repeatedly, your system cannot handle the requests and so it becomes overloaded and does nothing in response.

In addition to these attributes, we can also look at the abstract application methods that come from the class. There are some specific rules that abstract classes must follow that carry over into dynamic polymorphism. These are that you cannot simply create or declare abstract methods outside of the abstract class itself. As a general rule, classes abide by the fact that once class itself is sealed, nothing about it can be inherited or changed. Abstract classes, however, do not fall into this category as they can never be declared sealed.

Another term associated with polymorphism is the virtual function. These are decided at the initial run time and do have the ability to be brought in and used within an inherited class. Both of these, abstract classes and virtual functions, have the unique and important ability for this purpose to morph into many different formats and appearances as they can inherit classes and not be affected by a seal. This allows for a multi-dimensional and constantly different potential for changes within your programming and coding—something that is an excellent feature in gameplay programming.

Day 4- Interfaces and Namespaces

As you are creating something like a graphical interface, you obviously want to have the best quality you can produce with the most effective procedures possible. Interfaces consist of the large task of defining the members by deriving them from the class. You can look at the interface as the what portion or the syntactical contract and the deriving classes can be seen as the how. When it comes to using the interface to declare, you are going to see it in a similar way as class declaration. These statements are public by default as the interface itself is something that greatly interacts with the public.

With the interface you are able to compile and execute code that can tell you anything from the items you would see in a grocery store register to the graphical components of a game you were developing.

Just as with people names that we use in order to categorize and identify people, namespaces is a way to separate different names from each other. You formulate the namespace like so—

```
namespace namespace_name
{
  // code declarations
}
```

Seems simple enough, right? Just remember that, although this initially seems like a very simple concept, you are able to tack on different keywords and tasks within the realm of just the namespace. You can actually nest namespaces within each other simply by using . (dot) operator. Remember that namespace is different than class, but allows you to categorize names and separate them—an important distinction to be made from class!

Day 5- Preprocessor Directives and Regular Expressions

The basic function of the preprocessor is to give instructions that allow for a compiler to preprocess information before executing the actual compilation. The preprocessor always begins with # and another interesting and good fact about them to note is that they are not statements. Therefore, ending with a semicolon is never something that you do. Although the actual C# compiler is only consistent of the compiler itself, without a preprocess component separate; when one has a directive they are actually processed in a manner that would lead one to believe there is one working. A preprocessor is different than in other languages like C or C++ in C#, within the boundaries of this language, the instructor for the preprocessor must be the only one in the line.

Let's look at the possible directives that can be used in C# for the purposes of the preprocessor. We have #warning, #endregion, #error, #region, #line, #undef, #if, #else, #elif, #emdif, #line and finally #define. An interesting note about the @define preprocessor is that you can use it to create a symbolic constant that will, if is it deemed true, be passed to the #if directive. This will look like the following (once again in a general, all-purpose example format)—

#define symbol

From this now #if directive, we are able to use it to show and come up with a conditional directive that you can test different symbols for in order to determine their truth. A condition directive is to be formatted using the following—

#if symbol [operator symbol]...

Where the symbol in this line would represent what you would like to test in terms of a specific symbol. You can use some of our operator symbols (==, !=, && and ||) in order to evaluate the symbol itself within these parentheses. You also can assign these conditional directives in order to debug code or to give it specific configurations.

When you are looking at the input text, you can also match it up with a regular expression. These all have certain constructs that serve to define the expressions. These are anchors, character escapes, grouping constructs, character classes, backreference constructs, quantifiers, alternation constructs, miscellaneous constructs or substitutions. Figure out what you are looking to accomplish through your expressions and seek to find the appropriate boundaries (constructs) to put on them in order to produce the results you are looking for. Play around with these concepts for a while and figure out how to manipulate them.

Day 6- File I/O

So a file is something generally everyone is familiar with as a way that data is stored and the manner in which it is also stored (or the path). A stream is what occurs when you open a given file for either reading or writing. The two main streams are input and output that allow for data to be input or read and then the output is for writing into the given file.

There are quite a few different classes within this I/O (input/output) class. These are BinaryReader, BinaryWriter, BufferedStream, Director, DirectoryInfo, DriveInfo, File, FileInfo, Path, MemoryStream, StreamReader, StreamWriter, StringReader and StringWriter. You will use all of these types of file classes in order to build a namespace that allows for writing to, reading from and closing different files within your code. To create an object FireStream, you will need to use the following syntax—

FileStream <object_name> = new FileStream(<file_name>, <FileMode Enumerator>, <FileAccess Enumerator>, <FileShare Enumerator>);

Using the different parameters of FileMode, FileAccess and FileShare, you further define the class into the terms in which you see fit for the type of files you have.

Play with these concepts a lot, as they are important to really have a full and thorough grasp on in order to complete any projects you are working on. Remember to check for things like spacing and that you are placing things in the correct order. The power of one simple spacing or capitalization error could be what is throwing off your coding from producing the results you want.

Day 7- Review and Celebration

You have done it! You made it to the 21st day! On this day, much like the day of rest, you will want to go over your skills and make sure you feel confident in how all of the pieces put together make up the sum of C#.

Anything that irked you or caused you a multitude of mistakes, I would encourage you to go over again. I challenge you to go back to the first day and do an exercise that helps you to solidify concepts from each day that you have worked for the past 3 weeks to learn C#.

By reviewing and going over even just one concept from every day, you are stimulating those neural pathways that you have created. Every time you activate and use the pathway, it gets stronger. The more you practice and try our your coding and programming skills, the better and more confident you will become. Use this day as your final trial day before embarking on a project you feel passionate about. Still use your ability to try out anything you are building within a test environment before making it live; however, get ready to see the hard work and effort you have put forth into learning turn into action! The last chapter in this book is filled will resources that are useful for additional learning or to brush up on concepts that may be a little murky. As part of you final day, please read this final chapter during your celebration of all the hard work you have done!

Chapter 5- Helpful Hints and Resources

Although I believe that if you follow the steps of this book within the 21 day time period you will have a very thorough and firm grasp on C#, there are undoubtedly still questions you may have that have not been addressed. The wonderful thing about the technologically savvy world that we live in today is that this book is not your only resource to learning and mastering the programming and coding of C#. As I believe you now feel ready and able to take on many challenges you wish to address with your coding in the C# format, you may come to a crossroads where you are unsure of what the next potential step in problem solving is. This chapter should serve as your guide to finding any answers to questions that you have not received within the pages of this book.

First and foremost, I encourage you to look through your code for simple errors in spacing or capitalization. Sometimes if you are used to coding in a different language, you may have accidently put a period where a semicolon should be in C# or even put brackets where they should not be. These are the types of errors that I find to be most frustrating and typically are the mistakes that I have made. Essentially, always double check your work when you get to a frustrating point that seems to keep giving you an error message. Also, if you try this tactic and it doesn't work and you are ready to give up—take a break! Come back to your project after you have let yourself relax and take your mind off your project. When you come back and look at it with refreshed eyes you may find the error of your ways.

A resource that I utilize as well is online forums for C# in particular. Although some programming forums offer great general advice, I encourage you to google things like "C# help forum" in order to find others with potentially sage advice on how to solve an issue. Chances are you are not the only person to ever come across a certain speed hump that may be preventing you from completing a portion of your project. These can be great and accessible ways to quickly find an answer by using the power of the internet.

Although I already mentioned these next resources, I want to reiterate just how important that code editors and libraries can potentially be. Just as within any programming language, you can find seemingly endless answers in the libraries that will allow you to code for many different situations. As I suggested before, however, please have a firm grasp on the process of coding and programming (that this 21 day guide has given you!) before you venture into the world of using code. If something changes within the code and you can't find the mistake, you could potentially have irreparable damage to you program or project. Code editors can help you find these mistakes before making your code live, so I highly suggest that you consistently utilize this resource until you are a C# master—maybe even then just in case you fall victim to human error.

If you find that you need a better visual for any of the content within this book, or outside of the scope—I encourage you to visit my favorite video site—YouTube! Just search for a key phrase or word that you are attempting to figure out during your coding and programming and I guarantee that you will find a helpful video. Sometimes it may not be the very first video that you come across but it certainly will be on there. You may already have a preferred YouTuber that you go to for examples in a different language—maybe go look at the other users he or she subscribes to and see if their content can be helpful to you.

Never think that just because you are struggling it is the end of your project as a whole. I can tell you from personal experience that allowing yourself to fully explore and problem solve when issues like this come up, you are allowing yourself an opportunity to create new pathways in your brain that will serve you well with your struggles to come. Think back on the ways that you have solved other issues in your life, or programming experiences, and allow yourself to strategize how and why you can apply it to the situation at hand. You are your own best ally when it comes to knowing how you think and what you want to accomplish. Use your resources to the best of your ability and become as knowledgeable as you can—the power of knowledge and learning is quite a remarkable skill.

Programming and coding can be seen as a problem solving matter that you use to create programs that perform a specific purpose that can make things easier for you and the lives of others. Make sure that you are always working towards an achievable goal for your skill set, while continuing to push yourself to learn more and get to the next level of your functioning.

Conclusion

Now that this book has come to a close, you have all the information and knowledge of skills in order to be an excellent C# programmer and coder. It is important to me that you use this book as the foundation for your skills. I have worked hard to provide for you what I think is a guide that is comprehensive and easy to understand. Without languages like C#, we could not have the amazing technology both in our personal and business lives.

When you think about the opportunities that technological skills such a programming create in your life, I believe it is an invaluable skill. Knowing about the very basics of the computer system set up is something that can open doors for you in a professional sense, with more complex and high paying jobs. However, in your personal life, you can use the critical thinking and problem solving skills to help enhance your own gaming experience on your computer or other fields outside of technology. With the quickness you learned C#, you should feel very confident and fulfilled in your accomplishment!

Not everyone can learn an entirely new language in 21 days, which you have just simply and easily done through exploring a new virtual world. It is an accomplishment that you be happy about and really start to create the projects that you desired to. As an object oriented language, I am sure you have some excellent new ideas for contributions within the software community. Or perhaps you just want to be able to manipulate the Windows computer you have!

The descriptions and sections that I have broken down in this book should have allowed you to grasp and connect different pieces and parts of C# in a way that, when you reached this final conclusion, you are able to feel as though you can pass along knowledge to someone else. There are certain aspects of this language that I hope have reiterated my initial statements about this language being one that encourages gameplay programming. Through the many different avenues used in this language to define and separate different aspects that may touch (classes and statements for example) but are not interchangeable. It may seem like a bit of an overwhelming confusion as you get used to the nuances— the more you test out your skills and put examples of code to the test, the more you will come to realize the full impact and advantages of what these offer.

Thank you for picking up my book and giving it a chance to teach you something new! I would like to make every topic something that any person can learn the basics of quickly and easily. I hope you use this as a reference guide for your future projects and have been able to absorb the material and concepts with the help of the examples within the book. I want to end the book with a few more pearls of wisdom I find important to reiterate from my own experiences.

Always test your projects in a test zone before putting them life, especially if irrevocable damage could be done! Remember that you should be enjoying yourself when you are programming and coding because you are creating something complex by putting the pieces together yourself. Last, have fun and enjoy yourself while doing it!

www.ingramcontent.com/pod-product-compliance
Lightning Source LLC
Chambersburg PA
CBHW070719210526
45170CB00021B/872

9781542709095